NIGHT FLIGHT

SMG
SUNDANCE MEDIA GROUP

Night Flight Training Manual

SMG

SUNDANCE MEDIA GROUP

TABLE OF CONTENTS

Yuneec H520 with FoxFury Rugo lights are a flexible light source for night flight.

Introduction to Night Flight

Night flights are often some of the best any pilot can experience. The winds are usually calmer, seeing the city lights from above evokes emotions difficult to explain, and everything takes on an entirely new look. The view out the window or in the controller display is nothing short of spectacular, as even well-known sights take on a whole new look after sunset.

No matter what we might think of night flight, most RPC/Part 107 pilots (and general aviators) are well out of their element. Empirical data from various sources back this up, suggesting that many UAS operations at night end with some sort of minor incident, while a few are not so "minor."

Throughout these pages, we'll examine the topic in great depth, identifying best practices, concerns, emergency procedures, regulations, and equipment requirements.

What is the Definition of "Night Flight"?

> Night Flying is any flight that takes place 30 minutes *after* civil twilight, 30 minutes *prior* to before civil sunrise, or at any time between.

Can I Legally Fly UAS At Night?

The short answer is "Yes." The long answer is relevant to the sort of flight.

If the flight is for hobby/fun, and no commercial use is part of the process, night flight is possible under the rules of a

Community-Based Organization (CBO) such as the AMA. More on this can be found in the FARS, §Part 101 Subpart E: 101.41.

AMA rules for hobbyist flight state:

> *RC night flying requires a lighting system providing the pilot with a clear view of the model's attitude and orientation at all times. Hand-held illumination systems are inadequate for night flying operations.*
>
> *9. The pilot of an RC model aircraft shall:*
> *a) Maintain control during the entire flight, maintaining visual contact without enhancement other than by corrective lenses prescribed for the pilot.*
>
> *b)Fly using the assistance of a camera or First-Person View (FPV) only in accordance with the procedures outlined in AMA Document #550.*
>
> *c) Fly using the assistance of autopilot or stabilization system only in accordance with the procedures outlined in AMA Document #560.*

333 exempt operators do have access to a waiver to FAR 91.209(a), but to date, only one exemption to this has been granted. With 333 exemptions falling by the wayside and being replaced with Part 107 operations, there isn't any point in diving deep into this operational exemption.

Under Part 107 (Commercial Flight), it's a bit different. Remote Pilot Certificants require a waiver to 107.29.

CFR § 107.29

a. No person may operate a small unmanned aircraft system during night.

b. No person may operate a small unmanned aircraft system during periods of civil twilight unless the small unmanned aircraft has lighted anti-collision lighting visible for at least 3 statute miles. The remote pilot in command may reduce the intensity of the anti-collision lighting if he or she determines that, because of operating conditions, it would be in the interest of safety to do so.

c. For purposes of paragraph (b) of this section, civil twilight refers to the following:

1. Except for Alaska, a period of time that begins 30 minutes *before* official sunrise *and ends at* official sunrise;

2. Except for Alaska, a period of time that begins at official sunset and ends 30 minutes after official sunset; and

3. In Alaska, the period of civil twilight as defined in the Air Almanac.

There are multiple methods of calculating civil twilight and sunset, and charts are published both on the Internet and in various pilot guides.

March 2017
Captiva, Florida

Sunday	Monday	Tuesday	Wednesday	Thursday	Friday	Saturday
			1 Sunrise: 6:53am Sunset: 6:30pm	2 Sunrise: 6:52am Sunset: 6:30pm	3 Sunrise: 6:51am Sunset: 6:31pm	4 Sunrise: 6:50am Sunset: 6:31pm
5 ◗ Sunrise: 6:49am Sunset: 6:32pm First Qtr: 6:33am	6 Sunrise: 6:48am Sunset: 6:32pm	7 Sunrise: 6:47am Sunset: 6:33pm	8 Sunrise: 6:46am Sunset: 6:33pm	9 Sunrise: 6:45am Sunset: 6:34pm	10 Sunrise: 6:44am Sunset: 6:35pm	11 Sunrise: 6:43am Sunset: 6:35pm
12 DST Begins ○ Sunrise: 7:41am Sunset: 7:36pm Full Moon: 10:55am	13 Sunrise: 7:40am Sunset: 7:36pm	14 Sunrise: 7:39am Sunset: 7:37pm	15 Sunrise: 7:38am Sunset: 7:37pm	16 Sunrise: 7:37am Sunset: 7:38pm	17 Sunrise: 7:36am Sunset: 7:38pm	18 Sunrise: 7:35am Sunset: 7:39pm
19 Sunrise: 7:34am Sunset: 7:39pm	20 ◗ Sunrise: 7:33am Sunset: 7:40pm Last Qtr: 11:59am	21 Sunrise: 7:32am Sunset: 7:40pm	22 Sunrise: 7:31am Sunset: 7:41pm	23 Sunrise: 7:30am Sunset: 7:41pm	24 Sunrise: 7:29am Sunset: 7:42pm	25 Sunrise: 7:27am Sunset: 7:42pm
26 Sunrise: 7:26am Sunset: 7:43pm	27 ● Sunrise: 7:25am Sunset: 7:43pm New Moon: 10:58pm	28 Sunrise: 7:24am Sunset: 7:44pm	29 Sunrise: 7:23am Sunset: 7:44pm	30 Sunrise: 7:22am Sunset: 7:45pm	31 Sunrise: 7:21am Sunset: 7:45pm	

https://www.esrl.noaa.gov/gmd/grad/solcalc/ is one great resource, as is the Naval Observatory. These tables can also be downloaded from the Naval Observatory and customized for your location. The link for the Naval Observatory is http://aa.usno.navy.mil/publications/docs/aira.php). There are numerous charts on the web or in pilot handbooks. Sunrise/Sunset times are entirely based on geography; use a resource relevant to the local area of flight.

Night sUAS Operations Special Provisions. sUAS operations may be conducted at night, as defined in 14 CFR § 1.1, provided:

8. Prior to conducting operations that are the subject of this Waiver, the Responsible Person listed on the Waiver must ensure the remote PIC and VO are trained, as described in your petition for exemption, to recognize and overcome visual illusions caused by darkness, and understand physiological conditions which may degrade night vision. This training must be documented and must be presented for inspection upon request from the Administrator or an authorized representative;

9. The remote PIC and VO must ensure the area of operation is sufficiently illuminated to allow both the remote PIC and VO to identify people or obstacles on the ground, or the remote PIC and VO must conduct a daytime site assessment prior to conducting operations that are the subject of this Waiver, noting any hazards or obstructions;

10. In addition to the requirements of § 107.51(b), the sUA must not fly higher than 200 feet above any structure's immediate uppermost limit.

11. The remote PIC must ensure that all sUA remain inside the area of operation and that the remote PIC or VO can verify this at any time during operation.

TRAINING

The FAA requires training for night flight operations.

Training should include several hours of classroom time, flight time, and a test of theory and practical knowledge.

Training should define night flight regulations, equipment selection for night flight use, planning the night flight activity, safety considerations, briefings, and physical challenges associated with flying at night.

RULES OF NIGHT FLIGHT (FARS)

Refer to the FARs (Federal Aviation Regulations) Part 107.29 and AC107, in the Code of Federal Regulations (CFR).

CFR § 107.29

a. No person may operate a small unmanned aircraft system during night.

b. No person may operate a small unmanned aircraft system during periods of civil twilight unless the small unmanned aircraft has lighted anti-collision lighting visible for at least 3 statute miles. The remote pilot in command may reduce the intensity of the anti-collision lighting if he or she determines that, because of operating conditions, it would be in the interest of safety to do so."

AC-107 5.16

5.16 Daylight Operations. Part 107 prohibits operation of an sUAS at night, which is defined in part 1 as the time between the end of evening civil twilight and the beginning of morning civil twilight, as published in The Air Almanac, converted to local time. In the continental United States (CONUS), evening civil twilight is the period of sunset until 30 minutes after sunset and morning civil twilight is the period of 30 minutes prior to sunrise until sunrise. In Alaska, the definition of civil twilight differs and is described in The Air Almanac. The Air Almanac provides tables which are used to determine sunrise and sunset at various latitudes. These tables can also be downloaded from the Naval Observatory and customized for your location. The link for the Naval Observatory is http://aa.usno.navy.mil/publications/docs/aira.php.

5.16.1 Civil Twilight Operations

When sUAS operations are conducted during civil twilight, the small UA must be equipped with anticollision lights that are capable of being visible for at least 3 sm. However, the remote Pilot In Command (PIC) may reduce the visible distance of the lighting less than 3 sm during a given flight if he or she has determined that it would be in the interest of safety to do so, for example if it impacts his or her night vision. sUAS not operated during civil twilight are not required to be equipped with anti-collision lighting (We'll discuss anti-collision lighting in the next section).

This regulation may be waived with training and application.

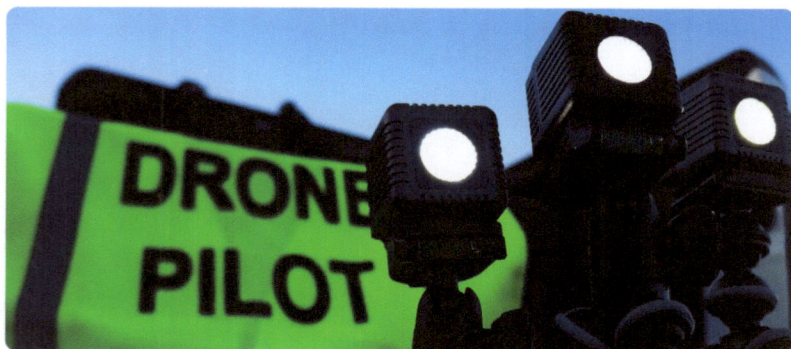

NIGHT FLIGHT PREPARATIONS

The UAS should have its own internal lighting system indicating front/back of the UAS. Although not required, it is very helpful to be able to identify front/back of the UAS.

Have fully-charged batteries both in the Ground Control Station/Remote, and for the UAS. Plan on flying a battery not more than 40% of capacity to allow for plenty of return to home time, landing time, and in the event of a lost UAS, that the battery will keep the UAS lit up in the event the UAS must be located through a search.

WHAT KIND OF LIGHTS DO I NEED FOR MY UAS?

This is where creativity comes into play.

No off-the-shelf UAS offers a lighting system that is compatible with FAA-regulated lighting requirements. A third-party lighting system will be required in order to meet regulation.

Night lighting requires a strobe light that strobes/blinks no slower than 40 cycles a minute, and not greater than 100 cycles per minute.

The lighting must be visible for three (3) statute miles.

Anti-collision lighting may be colored either red or white.

Adding non-OEM components may invalidate manufacturer authorizations. Further, non-OEM components may have an impact on weight, C/G, flight time, maneuverability, telemetry, IMU performance, or other negative affect on the overall performance of the UAS system.

ALWAYS test systems in daylight, and evaluate for the above issues. Always contact manufacturers for best recommendations regarding lighting systems to be sure that the system remains in compliance. While for hobbyists this practice may not be of great import, commercial operators, their insurance company, liability concerns, best industry practices may be issues requiring additional consideration.

LubeCube lights and mount on a Yuneec H520

THE NITTY-GRITTY DETAILS

FAR 23 provides guidance for those looking to specifically address (or DIY) lighting systems, and provides a clear understanding for manufacturers of anti-collision lighting for UAS systems. **Please note**: The regulation below is written for manned aircraft, but does apply (at this time) to UAS.

FAR § 23.1401 Anti-collision light system

a. General. The airplane must have an anticollision light system that:

1. Consists of one or more approved anticollision lights located so that their light will not impair the flight crew members' vision or detract from the conspicuity of the position lights; and

2. Meets the requirements of paragraphs (b) through (f) of this section.

b. Field of coverage. The system must consist of enough lights to illuminate the vital areas around the airplane, considering the physical configuration and flight characteristics of the airplane. The field of coverage must extend in each direction within at least 75 degrees above and 75 degrees below the horizontal plane of the airplane, except that there may be solid angles of obstructed visibility totaling not more than 0.5 steradians.

c. **Flashing characteristics.** The arrangement of the system, that is, the number of light sources, beam width, speed of rotation, and other characteristics, must give an effective flash frequency of not less than 40, nor more than 100, cycles per minute. The effective flash frequency is the frequency at which the airplane's complete anticollision light system is observed from a distance, and applies to each sector of light including any overlaps that exist when the system consists of more than one light source. In overlaps, flash frequencies may exceed 100, but not 180, cycles per minute.

d. **Color.** Each anticollision light must be either aviation red or aviation white and must meet the applicable requirements of § 23.1397.

e. **Light intensity.** The minimum light intensities in any vertical plane, measured with the red filter (if used) and expressed in terms of "effective" intensities, must meet the requirements of paragraph (f)of this section. The following relation must be assumed:

where:

I_e = effective intensity (candles)

$I(t)$ = instantaneous intensity as a function of time

$$I_e = \frac{\int_{t_1}^{t_2} I(t)dt}{0.2 + (t_2 - t_1)}$$

$t_2 - t_1$ = flash time interval (seconds)

Normally, the maximum value of effective intensity is obtained when t_2 and t_1 are chosen so that the effective intensity is equal to the instantaneous intensity at t_2 and t_1.

f. **Minimum effective intensities for anticollision lights.** Each anticollision light effective intensity must equal or exceed the applicable values in the following table.

Angle above or below the horizontal plane Effective intensity (candles)

0° to	5°	400
5° to	10°	240
10° to	20°	80
20° to	30°	40
30° to	75°	20

It should be noted here that in the Operation and Certification of Small Unmanned Aircraft Systems as published in the Federal Register, the FAA says; *"Position lighting may not be appropriate for some of these aircraft. Thus, instead of position lighting, small unmanned aircraft operating under part 107 will be required to have anti-collision lights when operating during civil twilight."* In short, positional lighting is not required for UAS night operations. **Only anti-collision lighting is required**. Therefore, although FAR 23.1401 provides direction for aircraft, only small portions apply to UAS.

Regardless of which lighting system is installed on the UAS, *always recalibrate the UAS' compass system prior to flight.* (Batteries in the lighting system, metal around the lighting housing, or other component may interfere with the UAS compass.)

Be certain to test lighting equipment in daylight! Installing lighting systems on a UAS may interfere with control or telemetry systems.

Lighting may interfere with camera systems; be certain to test/ evaluate lighting attached to the UAS for visual interference well prior to any working flight.

FLASHLIGHTS AND AREA LIGHTING

Obtain a red-light flashlight. Red Gel inserts are available for many flashlights. Do not attempt to use white light flashlights. White light will impact vision for minimum of 30 minutes (30 minutes of time are required for the eye to reacclimate to darkness). Red lighting helps mitigate the impact of light on the human eye.

If the operational area is to be lit, cover all lighting with red gel (available from Rosco or other theatre supply stores).

Even with red light, Visual Observers (V/O) and PIC's should never look directly at light. If the lighting faces the V/O's or pilot's field of vision, take care to always look to the side of the lighting instrument to prevent the eye from seeing the bright center of any lighting source.

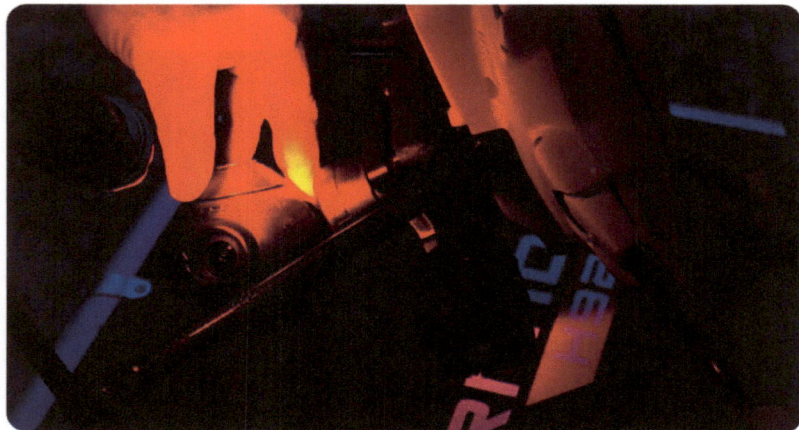

Red lighting helps provide protection for eye strain in night flight scenarios

If the tablet or viewing monitor does not offer a "night light" mode, cover the tablet or operational viewing screen with red gel to reduce eye strain and avoid as much parallax vision acuity loss as possible. iOS devices offer a "NightShift" mode, and Android devices offer a "Night" mode. Older Android systems can take advantage of various apps that will provide night time or twilight modes.

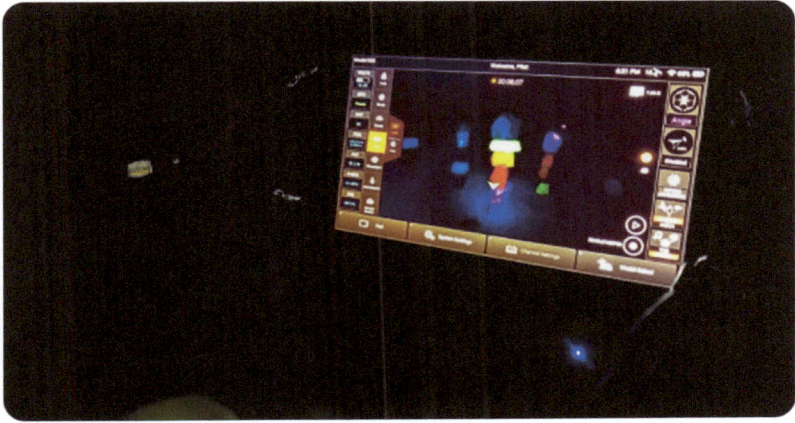

This image is covered in red gel (simulated) on one side to reduce eye fatigue and maintain best-possible vision

SCOUTING THE AREA IN DAYLIGHT

Whenever possible, always pre-flight the area in daylight. Take measurements to the target area if possible. Take note of any obstacles that may interfere with flight. Powerlines in particular, create an invisible hazard. Note the location, distance, and height of any powerlines, guy lines, phone lines, or other sorts of communication cables. These are invisible at night, and are frequently difficult to see during the day.

If walking the area isn't possible, Google Earth, or other mapping tools may help identify distances to objects/target areas. Knowing the distance between the control point and the target provides a reference, as distances are virtually impossible to accurately judge in the dark, particularly when viewing a UAS flying against a dark background. With reference distances, the pilot is able to refer to the distance between the control station and the UAS, and can easily calculate buffer distance for most efficient and safe operation.

In absence of a pre-flight measurement, another alternative is to place a V/O at the target, communicating with the PIC via radio.

Google Earth and other mapping tools are valuable in pre-planning night flights.

ENVIRONMENT/WEATHER

Weather is of course, a valid concern for night flight. Previously mentioned, winds are different at night. Although usually more calm, winds should be checked prior to any night flight. Cooling water, urban areas, desert sands may cause winds to arise at any point. **Check the METARs.**

Consider clouds. Clouds may diffuse moonlight and, while this provides a benefit of causing the UAS to be more visible against the diffused sky, be aware of cloud/fog altitude. Refer to Part 107 for a refresher on flying in clouds/fog.

Check the moon! While it might seem overkill, checking the phase of the moon during preflight is 90 seconds well spent. The difference between a full moon and a new moon is dramatic; flying "into" the moon isn't terribly different than flying into the sun, and from a contrast perspective, is far more dramatic than flying in the flattened contrasts of daylight.

The fact that the FARs require higher weather minimums at night is a good clue that you should be more pessimistic about weather. The standards of the FARS provide argument for knowing the air mass you're flying in, and understanding the big picture. If those cloud bases are flat and there's no precipitation to bring down visibility, conditions are likely suitable. Know the trends.

PERSONAL/INDIVIDUAL PREPARATION

NIGHT VISION PROTECTION

Several things can be done to help with the dark adaptation process and to keep the eyes adapted to darkness. General aviation pilots and flight crews frequently wear sunglasses or ND glasses to prevent eye pollution to protect their vision.

If a night flight is scheduled, pilots and crew members should wear neutral density (N-15) sunglasses or equivalent filter lenses when exposed to bright sunlight. This precaution increases the rate of dark adaptation at night and improves night visual sensitivity.

> **Be rested. Night flight is stressful, and being tired doesn't help.**

Remember to observe the "**8 hours bottle to throttle**" rule of aviation. No alcohol should be consumed at any point prior to flight. Even a single beer coupled with the inherent disorientation of flying in the dark, can cause an incident.

DO NOT DRINK AND FLY

CHALLENGES OF NIGHT FLIGHT

Challenges of night flight go beyond sight and ability to determine the area. The UAS and related equipment present their own set of challenges.

FLIGHT TIME (REDUCED)

- Lighting will add weight to the UAS. In most situations, this will reduce flight time. Plan for shorter/reduced flight time.
- Lighting adds drag to most UAS. Plan for shorter/reduced flight time.
- Lighting equipment that is not self-powered (powered from the UAS system) will be a drain on the flight battery. Plan for shorter/reduced flight time.
- The air is frequently more cool at night. This may have an impact on not only battery life, but also on UAS responsiveness.
- Check with the manufacturer of the UAS to provide direction for types of lighting. Remember, all FAA regulation falls back on "Manufacturer Recommendations."

MOUNTING SECURITY

This is where "home builds" may have great challenges, and if the system steps beyond products that have been evaluated by an aviation authority or manufacturer's recommendation, third-party or home-built lighting systems may be cause for an FAA action. For example, imagine flying over a car on the street and the method of attachment isn't secure. A light falls from the UAS, doing damage to a windshield or worse. This event could turn into an FAA action. Manufacturers of mounting systems should have primary and secondary mounting solutions so that in the case of a mount failure, there is still a secure system to prevent falling objects from injuring persons or damaging property. Of course, UAS should not be flying over persons not involved in the production while at the same time, motion in any direction at any speed may generate a trajectory that throws a lighting instrument beyond straight down. General aviators/pilots may not "just attach" new parts to manned aircraft without manufacturer guidance. Even in this new industry of unmanned, pilots should follow manufacturer recommendations.

> **Accident rates during night VFR decreases by nearly 50% once a pilot obtains 100 hours of flight experience and continues to decrease until the 1,000 hour level.**

MANEUVERING

The UAS will fly differently with added weight. Depending on the type of airframe, it's entirely possible that a sharp maneuver may cause the UAS to loose control or stability, and in the dark, it's entirely possible the pilot will not be able to immediately determine the nature of the instability.

Examining National Transportation Safety Board (NTSB) reports and other accident research can help a pilot learn to assess risk more effectively. For example, the accident rate during night visual flight rules (VFR) decreases by nearly 50 percent once a pilot obtains 100 hours and continues to decrease until the 1,000 hour level. The data suggest that for the first 500 hours, pilots flying VFR at night might want to establish higher personal limitations than are required by the regulations and, if applicable, apply instrument flying skills in this environment. Granted, this is a manned-aviation statistic, yet the value of experience equally applies to UAS operation. Stick-time is very valuable, and makes a significant difference when it comes to stressful situations. This is where the acronyms of general aviation come in. "PAVE" and "I'M SAFE" are two common acronyms that apply to all aviation, but are even more important when undertaking flight at night.

FATIGUE

Dealing with fatigue is primarily relevant to discipline. Two decade-old studies and reports all indicate a similar conclusion: there is no miracle cure for fatigue. We are not machines and after a full day of work, our attention and motor functionality slips. Sure, you can take caffeine, but it takes gallons of it to do

PAVE CHECKLIST	I'M SAFE CHECKLIST
PILOT 　Experience 　Currency **A**IRCRAFT 　Battery power/fuel 　Environmental 　performance 　Personal experience with 　this aircraft En**V**ironment 　Weather conditions (wind, 　rain, heat, cold) **E**xternal Pressures 　Power lines 　Guy wires	**I**LLNESS -Symptoms **M**EDICATION -Prescription or OTC **S**TRESS -Job, Health, Family Environment **A**LCOHOL -8 hours, 12 hours, 24 hours? **F**ATIGUE -Adequately rested, physically relaxed **E**ATING -Adequately nourished **E**MOTION-I feel mentally balanced and centered.

any good. The US Marine Corp has a deep study that demonstrates copious amounts of caffeine does nothing to improve motor function while denying sleep.

[Some of the following sections are adapted from the Pilot's Handbook of Knowledge, available on the FAA website at no cost.]

Be physically tuned for flight into reduced visibility. Ensure proper rest, adequate diet, and, if flying at night, allow for night adaptation. Remember that illness, medication, alcohol, fatigue, sleep loss, and mild hypoxia are likely to increase susceptibility to spatial disorientation.

VISION IN FLIGHT

Of all the senses, vision is the most important for safe flight. Most of the things perceived while flying are visual or heavily supplemented by vision. As remarkable and vital as it is, vision is subject to limitations, such as illusions and blind spots. The more a pilot understands about the eyes and how they function, the easier it is to use vision effectively and compensate for potential problems.

UNDERSTANDING HOW YOUR EYES FUNCTION

The eye functions much like a camera. Its structure includes an aperture, a lens, a mechanism for focusing, and a surface for registering images. Light enters through the cornea at the front of the eyeball (diagram follows)' travels through the lens, and falls on the retina. The retina contains light sensitive cells that convert light energy into electrical impulses that travel through nerves to the brain.

The brain interprets the electrical signals to form images. There are two kinds of light-sensitive cells in the eyes: rods and cones.

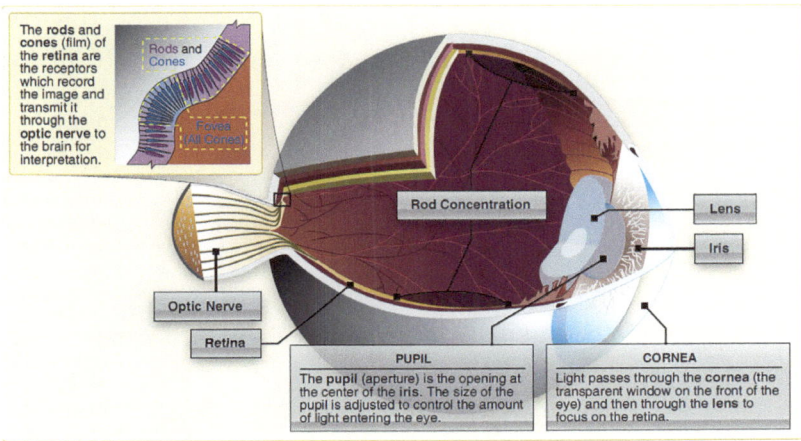

The rods and cones (film) of the retina are the receptors which record the image and transmit it through the optic nerve to the brain for interpretation.

Rods and Cones

Fovea (All Cones)

Rod Concentration

Lens

Iris

Optic Nerve

Retina

PUPIL

The pupil (aperture) is the opening at the center of the iris. The size of the pupil is adjusted to control the amount of light entering the eye.

CORNEA

Light passes through the cornea (the transparent window on the front of the eye) and then through the lens to focus on the retina.

The cones are responsible for all color vision, from appreciating a glorious sunset to discerning the subtle shades in a fine painting. Cones are present throughout the retina, but are concentrated toward the center of the field of vision at the back of the retina. There is a small pit called the fovea where almost all the light sensing cells are cones. This is the area where most "looking" occurs (the center of the visual field where detail, color sensitivity, and resolution are highest).

The cones are responsible for all color vision, from appreciating a glorious sunset to discerning the subtle shades in a fine painting. Cones are present throughout the retina, but are concentrated toward the center of the field of vision at the back of the retina. There is a small pit called the fovea where almost all the light sensing cells are cones. This is the area where most "looking" occurs (the center of the visual field where detail, color sensitivity, and resolution are highest).

While the *cones* and their associated nerves are well suited to detecting fine detail and color in high light levels, the *rods* are better able to detect movement and provide vision in dim light. The rods are unable to discern color but are *very sensitive at low-light levels.*

> As cone sensitivity decreases, pilots should use off-center vision and proper scanning techniques to detect objects during low-light levels.

The trouble with rods is that a large amount of light overwhelms them and they take longer to "reset" and adapt to the dark again. There are so many cones in the fovea that are at the very center of the visual field but virtually has no rods at all. So in low light, the middle of the visual field is not very sensitive, but farther from the fovea, the rods are more numerous and provide the major portion of night vision.

Vision Types

There are three types of vision: photopic, mesopic, and scotopic. Each type functions under different sensory stimuli or ambient light conditions.

Photopic Vision

Photopic vision provides the capability for seeing color and resolving fine detail (20/20 or better), but it functions only in good illumination. Photopic vision is experienced during daylight or when a high level of artificial illumination exists.

The cones concentrated in the fovea centralis of the eye are primarily responsible for vision in bright light. Because of the high light level, rhodopsin, which is a biological pigment of the retina that is responsible for both the formation of the photoreceptor cells and the first events in the perception of light, is bleached out causing the rod cells to become less effective.

Mesopic Vision

Mesopic vision is achieved by a combination of rods and cones and is experienced at dawn, dusk, and during full moonlight. Visual acuity steadily decreases as available light decreases and color perception changes because the cones become less effective. Mesopic viewing period is considered the most dangerous period for viewing.

Scotopic Vision

Scotopic vision is experienced under low-light levels and the cones become ineffective, resulting in poor resolution of detail. Visual acuity decreases to 20/200 or less and enables a person to see only objects the size of or larger than the big "E" on visual acuity testing charts from 20 feet away. In other words, a person must stand at 20 feet to see what can normally be seen at 200 feet under daylight conditions. When using scotopic vision, color perception is lost and a night blind spot in the central field of view appears at low light levels when the cone-cell sensitivity is lost.

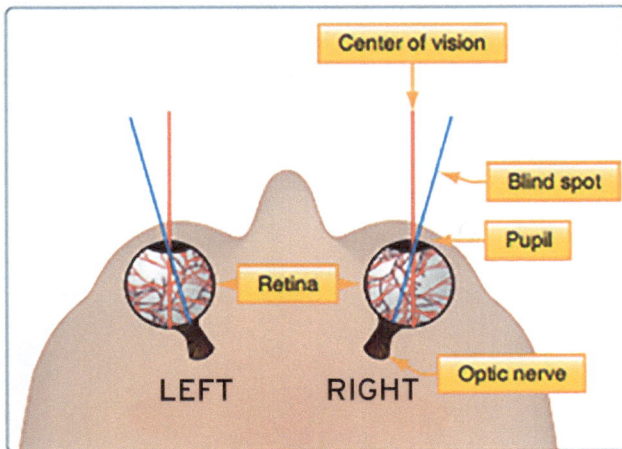

CENTRAL BLIND SPOT

The area where the optic nerve connects to the retina in the back of each eye is known as the optic disk. There is a total absence of cones and rods in this area, and consequently, each eye is completely blind in this spot.

As a result, it is referred to as the *blind spot* that everyone has in each eye. Under normal binocular vision conditions (both eyes used together), this is not a problem because an object cannot be in the blind spot of both eyes at the same time. On the other hand, where the field of vision of one eye is obstructed by an object (windshield divider or another aircraft), a visual target could fall in the blind spot of the other eye and remain undetected.

Finding Your Blind Spot

1. Hold this page at an arm's length.

2. Completely cover your left eye (without closing or pressing on it) using your hand or other flat object.

3. With your right eye, stare directly at the airplane on the left side of the picture page. In your periphery, you will notice the black X on the right side of the picture.

4. Slowly move the page closer to you while continuing to stare at the airplane.

5. When the page is about 16–18 inches from you, the black X should disappear completely because it has been imaged onto the blind spot of your right eye. (Resist the temptation to move your right eye while the black X is gone or else it reappears. Keep staring at the airplane.)

6. As you continue to look at the airplane, keep moving the page closer to you a few more inches, and the black X will come back into view.

7. There is an interval where you are able to move the page a few inches backward and forward, and the black X will be gone. This demonstrates to you the extent of your blind spot.

8. You can try the same thing again, except this time with your right eye covered stare at the black X with your left eye. Move the page in closer and the airplane will disappear.

NIGHT BLIND SPOT

It is estimated that once fully adapted to darkness, the rods are 10,000 times more sensitive to light than the cones, making them the primary receptors for night vision. Since the cones are concentrated near the fovea, the rods are also responsible for much of the peripheral vision. The concentration of cones in the fovea can make a night blind spot in the center of the field of vision.

To see an object clearly at night, the pilot must expose the rods to the image. This can be done by looking 5° to 10° off center of the object to be seen. This can be tried in a dim light in a darkened room. When looking directly at the light, it dims or disappears altogether. When looking slightly off center, it becomes clearer

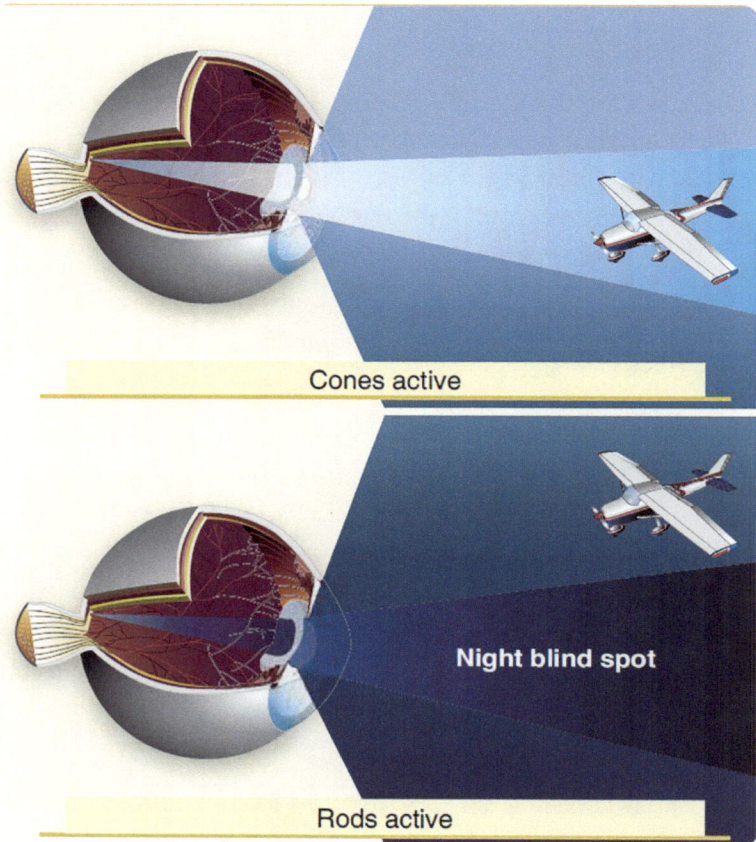

Cones active

Night blind spot

Rods active

and brighter. When looking directly at an object, the image is focused mainly on the fovea, where detail is best seen.

At night, the ability to see an object in the center of the visual field is reduced as the cones lose much of their sensitivity and the rods become more sensitive. Looking off center can help compensate for this night blind spot. Along with the loss of sharpness (acuity) and color at night, depth perception and judgment of size may be lost.

DARK ADAPTATION

Dark adaptation is the adjustment of the human eye to a dark environment. That adjustment takes longer depending on the amount of light in the environment that a person has just left. Moving from a bright room into a dark one takes longer than moving from a dim room and going into a dark one. While the cones adapt rapidly to changes in light intensities, the rods take much longer. Walking from bright sunlight into a dark movie theater is an example of this dark adaptation period experience. The rods can take approximately 30 minutes to fully adapt to darkness. A bright light, however, can completely destroy night adaptation, leaving night vision severely compromised while the adaptation process is repeated.

SCANNING TECHNIQUES

Scanning techniques are very important in identifying objects at night. To scan effectively, pilots must look from right to left or left to right. They should begin scanning at the greatest distance an object can be perceived (top) and move inward toward the

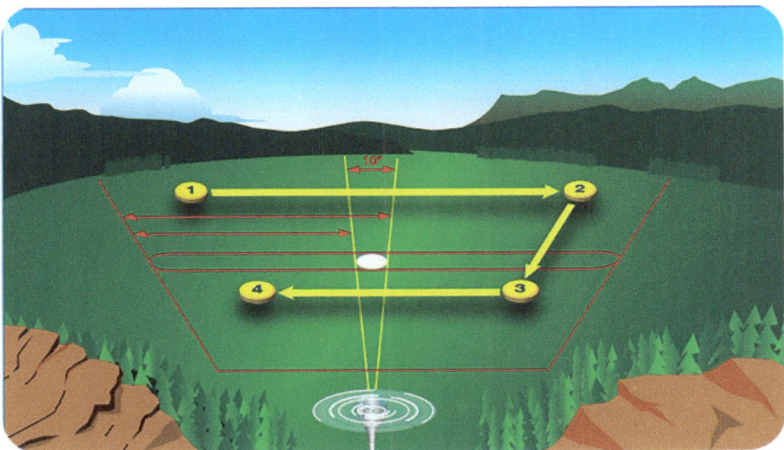

position of the aircraft (bottom). For each stop, an area approximately 30° wide should be scanned. The duration of each stop is based on the degree of detail that is required, but no stop should last longer than 2 to 3 seconds. When moving from one viewing point to the next, pilots should overlap the previous field of view by 10°.

Off-center viewing is another type of scan that pilots can use during night flying. It is a technique that requires an object be viewed by looking 10° above, below, or to either side of the object $_{(diagram\ Pg.\ 35)}$. In this manner, the peripheral vision can maintain contact with an object. With off-center vision, the images of an object viewed longer than 2 to 3 seconds will disappear. This occurs because the rods reach a photochemical equilibrium that prevents any further response until the scene changes. This produces a potentially unsafe operating condition.

> To overcome this night vision limitation, pilots must be aware of the phenomenon and *avoid* viewing an object for longer than 2 or 3 seconds. The peripheral field of vision will continue to pick up the object when the eyes are shifted from one off-center point to another.

NIGHT VISION PROTECTION

Several things can be done to help with the dark adaptation process and to keep the eyes adapted to darkness. Some of the steps pilots and flight crews can take to protect their night vision are described in the following paragraphs.

SUNGLASSES

If a night flight is scheduled, RPC pilots and Visual Observers should wear neutral density (N-15) sunglasses or equivalent filter lenses when exposed to bright sunlight. This precaution increases the rate of dark adaptation at night and improves night visual sensitivity.

FLIGHTDECK LIGHTING

Flightdeck lighting (remote/groundstation lighting) should be kept as low as possible so that the light does not monopolize night vision. After reaching the desired flight altitude, pilots should allow time to adjust to the flight conditions. This includes readjustment of instrument lights and orientation to outside references. During the adjustment period, night vision should continue to improve until optimum night adaptation is achieved. When it is necessary to read maps, charts, and checklists, use a red light flashlight and avoid shining it in your or any other crew members eyes.

The UAS launch area lighting should be reduced to the lowest usable intensity. All onsite crew should practice light discipline with headlights and flashlights. Position the UAS at a part of the airfield where the least amount of lighting exists.

SELF-IMPOSED STRESS

Night flight can be more fatiguing and stressful than day flight, and many self- imposed stressors such as tension, can limit night vision. Pilots can control this type of stress by knowing the factors that can cause self-imposed stressors.

DISTANCE ESTIMATION AND DEPTH PERCEPTION

Knowledge of the mechanisms and cues affecting distance estimation and depth perception assist pilots in judging distances at night. These cues may be monocular or binocular. The monocular cues that aid in distance estimation and depth perception include motion parallax, geometric perspective, retinal image size, and aerial perspective.

MOTION PARALLAX

Motion parallax does not apply to UAS flight. Motion Parallax applies only to manned aviation, as to experience Motion Parallax, one must be inside the moving vehicle or in a simulated environment.

GEOMETRIC PERSPECTIVE

An object may appear to have a different shape when viewed at varying distances and from different angles. Geometric perspective cues include *linear perspective, apparent foreshortening,* and *vertical position in the field.*

Linear Perspective—parallel lines, such as power lines and railroad tracks tend to converge as distance from the pilot or visual observer increases. This creates the illusion of a "vanishing point" and can be seen in a tablet or viewing monitor.

Apparent Foreshortening—the true shape of an object or a terrain feature appears elliptical when viewed from a distance.

Vertical Position In The Field—objects or terrain features farther away from the observer appear higher on the horizon than those closer to the observer.

NIGHT VISION IN SUMMARY

Knowledge of the basic anatomy and physiology of the eye is helpful in the study of night operations. A knowledge of visual illusions provides the remote pilot methods to overcome those illusions. Techniques for preflight and night approach help teach the remote pilot safer ways to conduct flight at night.

> # Safety should ALWAYS be the primary consideration.

OTHER BASIC SAFETY/RISK MITIGATION PRACTICES AND ISSUES

There are several hazards in night flight. Some have been discussed previously and will be reiterated here.

A UAS pilot couldn't identify power lines in the monitor, even in daylight. Image courtesy Seattle Times.

Guywire, powerlines, telephone lines, and other overhead cabling are a significant risk. These cables absolutely cannot be seen during night flight. When possible, it is good practice to tie reflective tape to guy wires, and to mark out on the ground (with reflective material) where energy or communication lines exist. A V/O may be placed near/beneath energy or communication lines as an alternative.

Another night hazard are trees. Leaves, branches, and palms are difficult to see, and the same actions for energy or communication lines also apply to foliage.

Winds may be different at night, as the earth is cooling vs warming, and turbulence, different wind speed, wind direction may/ are likely to occur at night. Areas where rotors and turbulence may not have an impact during the day may be entirely different at night. This again, a strong argument for walking the area prior to sunset. An experienced pilot is able to identify potential object rotors or object turbulence regardless of wind direction.

CHOOSE AND IMPLEMENT RISK CONTROLS

Once you have perceived a hazard (step one) and processed its impact on flight safety (step two), it is time to move to the third step, perform.

Perform risk management by using the TEAM checklist: Transfer, Eliminate, Accept, Mitigate to deal with each factor.

Transfer—Should this risk decision be transferred to someone else (e.g., do you need to consult the with the V/O or other briefed team members?)

Eliminate—Is there a way to eliminate the hazard?

Accept—Do the benefits of accepting risk outweigh the costs?

Mitigate —What can you do to mitigate the risk?

NIGHTFLIGHT CHECKLIST

We recommend using the checklist on the next page as a basic tool prior to any night flight activity.

☑ Walk the area to be flown.
    o Note obstacles, discuss with Visual Observer
    o Be sure communication devices are functioning correctly & batteries are charged
    o Check lighting on object to be filmed in the darkness
☑ Receive an updated METAR or WX report
☑ Double check all ATC or airspace permissions (if necessary)
☑ Inspect area for potential sources of light pollution
☑ Verify all transmitter, on-board aircraft, lighting, and camera batteries are fully charged; (confirm voltages)
Checking all control surfaces for signs of damage, loose hinges, and overall condition;
    o Looking over the wing/rotors to ensure they are in good structural condition and properly secured;
☑ Check propellers / mounting hardware (tight) / rotor blades for chips and deformation;
☑ Inspect the landing gear for damage and function
☑ Ensure photo/video/lighting equipment mounting system is secure and operational.
☑ Check location of GPS equipment controlling the autopilot.
☑ Check the IMU movements in the ground control software.
☑ UAS in stabilization mode, ensure control surfaces move towards the correct positions
☑ UAS is in a level location safe for takeoff
☑ Camera settings are correct (still images, video, framerate)
☑ SD camera memory formatted and inserted into the camera
☑ All transmitter controls move freely in all directions
☑ All transmitter trims in neutral position
☑ All transmitter switches in correct position(typically centered or upward/away from pilot)
☑ Transmitter throttle to zero
☑ Ground station powered on
☑ Power UAS
☑ Ensure led indicators and audible tones are correct
☑ Scan for nearby cars / people / animals
☑ Call "CLEAR!" in a loud voice
☑ Arm flight controller/rotate props. Listen for any abnormalities
☑ Call "Launch" in a loud voice
☑ Launch UAS
    o Short 20-30 second hover at 7-10 feet (listen for vibrations / loose items)
☑   o Never at eye level
☑ Perform a controllability check (up/down/right/left/yaw right/yaw left)
☑ Confirm in Ground Control station that battery and GPS values are correct
☑ Fly the mission

Hoodman offers a landing pad and a separate/optional lighting system in green, ideal for night flight. The lighting system offers three modes: Static light, pulsing light, or blinking light.

Applying for a 107.29 Night Flight Waiver

Once training is complete, both theoretical and practical, it is time to file an application for a night flight waiver.

This can be achieved through visiting the FAA website page https://www.faa.gov/uas/request_waiver/request_part_107_waiver/

In this webpage, the FAA has several questions relevant to the operation procedures and backup contingency plans.

For each regulation subject to waiver that you checked above, please provide details about how you will meet the *waiver safety explanation guidelines*.

* Waiver safety explanation:

Character count: 0/5,000

You will receive a confirmation email once your waiver application has been in-processed by the FAA. Follow the instructions in this email to submit supporting documentation for your waiver request. Supporting documents must be provided within 7 days of receiving the confirmation email.

Be certain to fill out this section in your own words, not words copy/pasted from someone else' application.

Specifically describe how:

— The pilot will maintain visual line of sight during darkness/ night operations

— Provide explanation of how other aircraft will be observed and avoided, and how persons/property on the ground will be avoided. Identify how obstacles will be observed and avoided

— Explain how the pilot will be aware of position, altitude, attitude, speed, and movement of the UAS

— Demonstrate the all persons participating in the operation will have knowledge of, and ability to compensate for visual illusions caused by darkness, bright light, light pollution, and other physiological conditions that degrade night flight

— Explain how the pilot will not be distracted nor engaged in any other activity during night flight operations

— Demonstrate how the aircraft will be made visible/conspiculous for 3SM, or demonstrate why this is not necessary (remote areas)

— Demonstrate any relevant maps, obstacles, or areas that indicate awareness of hazards.

We suggest adding language that indicates how the aircraft may be located in the event of a failed flight/crash/battery failure. Our systems for example, use separate power for the beacon lights. In the event of a downed aircraft, the aircraft power system does not affect the power system of the strobe.

The majority of 107.29 waivers are rejected by the FAA for a variety of reasons, yet the greatest cause of rejected waivers falls into two categories;

1. **Incomplete waiver application**
2. **Copy/pasted waiver application and explanation of various procedures.**

The first of these issues is easily overcome through careful reading of the application, ensuring all areas of the application are properly completed. Leave no field empty; use "n/a" if the field is non-applicable to the specific application.

Be certain all boxes are checked.

Include specific details to the operation and all relevant aspects to compliancy, secondary plans, emergency plans, crew requirements/training, and any other relevant data. Describe the environment, involved persons, means of avoiding potential distractions, lost-link procedures, and safety precautions.

Be verbose/specific. Filling out the relevant section with "I'm going to execute a process that does XYZ" will not achieve the desired result. Express in great detail what the plan will entail. Being vague assures a failed/denied application. To reiterate; the FAA wants to know what you're doing, where you're doing it, and how you're going to be safe as to the specific situation that that flight is going to occur.

Be sure to explain how the ground crew will be trained to avoid light pollution, be aware of the physiological issues, and understand the criticality of awareness during non-daylight operations. Express how the crew will be made aware of ADM and risk management. This book is a good start.

DO NOT COPY/PASTE data from any FAA or online community site. The FAA will recognize a copy/paste application and almost certainly deny the application based on that this common error. The performance-based standards the FAA explains on their website are for thoughtful consideration, not copy/pasting.

Have someone review your application prior to submission. Spelling errors, poor grammar, or poorly described/unclear explanations may be cause for denial. Think of the blind men describing the elephant; concise and clear explanations for each execution will be valuable. Proofreading by another person may make the difference in an approved or denied application.

Be sure to include your Remote Pilot Certification number. If the permanent card has not yet been received, simply but "pending" in the application. Include your Commercial UAS registration number as well.

It is unlikely that too much information is a problem with a waiver application. The entire point of the waiver application is to demonstrate that the PIC is aware of the challenges, has responses/executions to obviate the challenges, and contingency plans in the event of mission failure.

DIFFERENT AIRSPACE REQUIREMENTS

Applying for the 107.29 waiver for Class G airspace is relatively cookie-cutter in how the FAA grants this waiver. The Class G waiver from the FAA generally is a long-term waiver, with some specific restrictions. However, this waiver will almost assuredly not apply to Class B, C, D, or E airspace, and those operations will likely require a specific application for a specific flight operation.

AFTER THE APPLICATION

What happens after the application is submitted?

The FAA will examine the application. This may take up to 90 days. Once the application is viewed by a human, one of three actions will occur:

1. **The waiver will be granted/approved, and the formal form of approval will be included with the notice of approval.**

2. **The application will be denied, with little or no explanation of reason for denial.**

3. **The FAA may ask for further information.**

If a response has not been received *within 90 days*, raise a hand; contact the FAA and inquire about the application.

A failed application looks like this:

Thank you for submitting a 14 CFR Part 107 request for certificate of waiver. In accordance with 14 CFR § 107.200, your application has been reviewed and found to have incomplete information, or the interventions you described are insufficient to mitigate risk to an acceptable level. We are currently unable to process your request any further. Please review the attached letter, and if you would like to reapply, provide the required information in your application. You can find the performance standards required at https://www.faa.gov/uas/request_waiver/.

Sincerely,

Part 107 Waiver Team
General Aviation and Commercial Division

U.S. Department
of Transportation
**Federal Aviation
Administration**

RE: 14 CFR Part 107 Waiver Reference Number ▮▮▮▮▮

Dear ▮▮▮▮▮

Thank you submitting a 14 CFR Part 107 request for certificate of waiver through the automated FAA small unmanned aircraft (sUAS) waiver application portal.

When the FAA responds to a request for a certificate of waiver, it must follow the requirements of 14 CFR § 107.200, "Waiver policy and requirements," particularly those standards outlined in § 107.200(b). As stated in that section, the FAA uses the following criteria when making a decision as to whether to grant a waiver:

1) a complete description of the proposed operation; and
2) justification that establishes that the operation can safely be conducted under the terms of a certificate of waiver.

Given the criteria outlined above, the FAA is unable to approve your request for a waiver to § 107.29 because you did not describe interventions, for one or more hazards, to mitigate risk to an acceptable level. If you would like to reapply, include as much detail as required to describe the proposed operation, the purpose of the operation, and method by which the proposed operation can be safely conducted. Information should identify potential hazards and risks of the waivered operation, including risk-mitigation strategies, and characteristics of the sUAS. Refer to the waiver performance standards at: https://www.faa.gov/uas/request_waiver/. You must address each of the standards for the applicable regulatory section to be waived. Address each standard and how you propose to mitigate risks associated with the hazards utilizing operating limitations, technology, training, equipment, personnel, sterile areas, etc. Only request waiver from regulatory sections necessary to conduct the operation.

Thank you for supporting the FAA in the safe integration of UAS operations into the National Airspace System.

Sincerely,

BY DIRECTION OF THE ADMINISTRATOR

Carl Johnson Digitally signed by Carl Johnson
DN: cn=Carl Johnson, o=General Aviation and Commercial Division, ou=AFS-800, email=carl.k.johnson@faa.gov, c=US
Date: 2016.08.28 19:55:59 -04'00'

General Aviation and Commercial Division, AFS-800

This is the only notification the FAA will provide in the event of a denied application

A successful application will include the Certificate of Waiver, and will appear as:

U.S. DEPARTMENT OF TRANSPORTATION
FEDERAL AVIATION ADMINISTRATION

CERTIFICATE OF WAIVER OR AUTHORIZATION

ISSUED TO

Douglas W. Spotted Eagle
Waiver Number: ▓▓▓▓▓▓▓

ADDRESS

▓▓▓▓▓▓▓

This certificate is issued for the operations specifically described hereinafter. No person shall conduct any operation pursuant to the authority of this certificate except in accordance with the standard and special provisions contained in this certificate, and such other requirements of the Federal Aviation Regulations not specifically waived by this certificate.

OPERATIONS AUTHORIZED

Night small unmanned aircraft system (sUAS) operations in accordance with petition for exemption, docket number FAA-2016-7611

LIST OF WAIVED REGULATIONS BY SECTION AND TITLE

14 CFR § 107.29 Daylight operation

STANDARD PROVISIONS

1. A copy of the application made for this certificate shall be attached to and become a part hereof.
2. This certificate shall be presented for inspection upon the request of any authorized representative of the Administrator of the Federal Aviation Administration, or of any State or municipal official charged with the duty of enforcing local laws or regulations.
3. The holder of this certificate shall be responsible for the strict observance of the terms and provisions contained herein.
4. This certificate is nontransferable.

NOTE—This certificate constitutes a waiver of those Federal rules or regulations specifically referred to above. It does not constitute a waiver of any State law or local ordinance.

SPECIAL PROVISIONS

Special Provisions Nos. 1 to 11, inclusive, are set forth on the attached pages.

This Certificate of Waiver 107W-2016-00130 is effective from August 29, 2016 to August 31, 2020 and is subject to cancellation at any time upon notice by the Administrator or an authorized representative.

BY DIRECTION OF THE ADMINISTRATOR

Carl Johnson

Digitally signed by Carl Johnson
DN: cn=Carl Johnson, o=General Aviation and Commercial
Division, ou=AFS-800, email=carl.n.johnson@faa.gov, c=US
Date: 2016.08.28 19:55:59 -04'00'

General Aviation and Commercial Division, AFS-800

FAA Form 7711-1 (7-74)   CC:

Keep this waiver on your person during all night flight operations. SMG recommends a policy and procedures manual accompany all operations (in printed form with access to an online version in the absence of a printed manual). This certificate is required to present upon request from any law enforcement or FAA official.

What To Do If My Application Is Denied?

A denial of application likely falls back on the two issues presented earlier in this section.

1. Incomplete waiver application
2. Copy/pasted waiver application and explanation of various procecures.

Begin the application process again, taking extra care to double or triple check all fields. It's quite possible something small, yet important was missed.

Ask a coworker or friend to review your application.

This may be a good opportunity to get to know the Aviation Safety Inspector(s) in your area. Contact your local Flight Services District Office (FSDO, frequently referred to as a "Fizz-dough") and ask for an appointment with the local sUAS ASI. Most offices have at least one assignee, while other offices may have more. Don't expect a same-day appointment. UAS ASI's have other duties in addition to their UAS duties. Think of them as "detectives" on a regular police force.

Ask the ASI where the application is flawed. Ask them to help you understand specific requirements that may have been missed in the first application. Understand that to them, the Remote Pilot Certificate is merely a "drivers license" and they do expect that the applicant has some basic knowledge. Be studied; have specific questions prepared in advance so that the time is efficiently used. Remember, their primary job is to investigate incidents, not train new pilots. However, it has been this authors experience that the ASI's are generally quite friendly and helpful, so long as their time and their person are treated with due respect.

Resubmit the application once all errors and/or omissions are corrected.

What About Using An Attorney To Apply?

There are many online attorneys offering to manage waiver submissions for a fee. Of course, they can be used for form submission, and will likely provide a positive result. However, if one is seeking a career in unmanned aviation, it is going to require submissions for waivers or airspace authorizations on a reasonably regular basis, particularly as the new FAA online waiver application process comes into place. This could become very expensive, but the greater point is that pilots need to understand the "whys" of the application process in order to better understand the "whats" of the application process and mission/flight execution. If an operation arises that requires some very unique permissions, such as night flight near a TFR over 400', then perhaps an attorney may be of benefit. Otherwise, we recommend pilots learn these processes and applications as a matter of course for their business operations.

About Sundance Media Group

Based in Las Vegas, Nevada and West Jordan, Utah, Sundance Media Group (SMG), has been producing training for trade events for nearly 20 years. Instructors from SMG have taught, presented workshops and have participated in panels worldwide. Over the years, SMG's area of focus has been audio, video, and software applications for production and post-production. Douglas Spotted Eagle, the original founder of SMG, has long history of aviation, from the adrenaline-filled world of fast-action videography in skydiving to commercial application of drone/UAV use. SMG's latest evolution and vision is to incorporate its years of experience for best-practices training into the world of UAV use.

SMG serves as a consultant within the UAV industry, offering training and speaking engagements on UAV topics ranging from, but not limited to: UAV cinematography, commercial and infrastructural UAV applications, UAV risk management, night UAV flight, aerial security systems, and 107 training to ensure pilots clearly understand the FAA laws. SMG has intimate knowledge of the FAA FARs and FSIM; our collective experience with instructors and UAV pilots nationwide is our foundation for creating a best-practices for everything drone/UAV/UAS. The greatest strengths in the SMG lineup of consulting and education services are the vertical-specific training programs for Public Safety, Journalists, Cinematographers, Thermal uses, and mapping/infrastructure development.

DOUGLAS SPOTTED EAGLE

Douglas Spotted Eagle is a giant in the video and audio industries, having received Grammy, Emmy, DuPont, Peabody, and many other awards. Douglas is a primary instructor and industry consultant for Sundance Media Group, Inc. and VASST, authoring several books and DVDs and serving as an advisor and guide for videography/software manufacturers and broadcasters. Douglas is a well-known musician and a world-travelled speaker/instructor.

Skydiving since 2006 and instructing UAS since 2012, Douglas is an accomplished aerial photographer who thrives in the adrenaline-filled world of fast-action videography. Appointed as a Safety and Training Advisor in the aviation world, he is a risk management/mitigation subject matter expert. Douglas is an audio and imaging pro with numerous awards for his productions; with an intimate knowledge of the FAA FARs and FSIMs, Douglas' vision is to incorporate his years of imaging and aviation experience into best-practices for Sundance Media Group's. Douglas is a frequent (and dynamic!) speaker and consults on many UAV subjects.

JENNIFER PIDGEN

As majority owner and COO of Sundance Media Group (SMG), Jennifer is dedicated to developing the sUAS/UAV training programs and strategic industry partnerships. A marketing guru with over 20 years of marketing experience within the consumer electronics and photo/ video channels, Jennifer also manages large-scale training events and vendor/sponsor relationships. No stranger to logistical and analytical reporting, Jennifer manages all sUAS/UAV logistics and overall SMG operations, including applying for SMG's ISO certification. Bearing a degree in finance/accounting, Jennifer also holds a USPA regional wingsuit and speed skydiving judge rating. Jennifer's diverse background in science, math, and marketing are a remarkable combination within the sUAS industry as her education, experience, entrepreneurship, and passions are put to good work. Ultimately, her expertise is inspiring conversation and cultivating mutually-beneficial partnerships; each with a focus towards building a successful and safe sUAS/UAV community.

WITH SPECIAL THANKS

For editing, insights, images and more, a special thank to our instructors Luisa Winters, James Spear and Jack Spear.

For keeping us on our toes with the "nitty-gritty" details, thank you to Ron Campbell of SGS HART Aviation.

www.ingramcontent.com/pod-product-compliance
Lightning Source LLC
Chambersburg PA
CBHW041226270326
41934CB00001B/12